Uncovering the Secrets of the Lymphatic System

The Possible Role of Sodium Protein Chains as the Transport Mechanism for the Lymphatics and Beyond

Kenneth W. Liskey Sr.

The Lymphatic Frequency Theory

Original Conception Date: 1/2/16

Revised Date: 8/02/19

A Case for Sodium Proteins as the Fundamental Transporter of Waste between the Lymph System and All Other Body Systems

ISBN: 978-0-578-55428-0

ken.liskey@gmail.com

Contents

Acknowledgement

I would like to acknowledge the Higher Power, for teaching me to embrace calm, during the turmoil of my life. Enabling me to remain still, so that I could open my eyes and glimpse these truths, hidden in plain sight.

Abstract

This book presents a theory to describe the unseen mechanism, which enables the body to transfer waste from the circulatory system, the nervous system and the interstitial fluid, towards and then through the lymph system. Where the directional fluid flow is the collective result of the pull-push pumping action of the individual sodium proteins, found in the interstitial fluid and lymph fluid, all working together in a frequency.

Hence, the interstitial/lymph fluid flow is actually generated from the ionic pull, created by the in-line sodium proteins chains. As such, this concept recognizes the gatekeeper status of the uni-directional lymph valves, whose priority it is to assist with the directional flow. But, minimizes the role played by the lymph valves and vessels in maintaining the constant head pressure required to achieve flow throughout the lymph system.

Accordingly, during a toxic waste event, the lymph system can still function, even when a node or valve becomes compromised. To do so, the theory suggests a series of deviations may transpire at the sodium protein level, which permits the body to temporarily circumvent the lymph system at the point of failure.

Eventually, these sodium protein reactions may also provide the body with the ability to override the underperforming one-way lymph valves, along with changing the direction of the uni-flow lymph fluid. Such a strategy to sidestep the lymph system or to back flow the contaminated fluid stagnated within the lymphatics, would serve the dual purpose of waste cleanup deferment and waste redistribution.

This predicted ability of the body to evade or back flow the lymphatics during an emergency, is a necessary survival strategy. But under chronic conditions such a deliberate waste management approach will eventually lead to certain types of cancer, circulatory and neurological disorders, lymph edema, arthritis and many other syndromes.

Fortunately, if we can identify and monitor the sodium protein changes occurring during a toxic waste event and learn to manipulate the lymphatic transport system at the sodium protein level, we will be able to front run many illnesses, by intervening prior to the onset of the disease.

The Lymphatic Frequency Theory

Physical connections are minimally discernable between the lymphatic system and the other body systems, including the circulatory system, the brain, the nervous system and the interstitial region. Yet, under normal conditions the lymph system can still clear waste successfully from these areas.

The ionic capabilities found within sodium proteins may aid the lymphatic system in removing unwanted toxins, pathogens and waste from these various areas, via the interstitial fluid and/or lymph fluid. This process initiates within the interstitial region, where free-floating sodium attaches to the undesirable waste material, and is then collected by sodium proteins, prior to entering the lymph system.

Once the contaminated sodium is captured within the protein, it is then pumped towards the lymph system, along with the surrounding fluid. Only to be recaptured downstream by another protein, so in essence, the in-line sodium proteins form a collective pumping chain, which moves the interstitial fluid and sodium towards the lymph vessels. It is this two-beat pump action of the sodium proteins that extends the influence of the lymphatic vessels, making it possible to draw the captured waste from the blood and surrounding tissues to the lymph system, without a physical connection.

After the captured toxins are transferred to the lymphatics, sodium proteins are then utilized by the lymph system for channel transportation within the lymph vessels, again due to their great ionic capacity and pump action. As described, such transference of toxins both outside and within the lymphatic system would be made possible, in part due to the sodium-potassium pump mechanism of the individual sodium proteins.

Of equal importance is the architecture of the lymphatic vessels, which start out smaller in the body's extremities, and grow in diameter with the direction of the lymph fluid flow. This arrangement of increasing vessel diameter, is similar to a plant's root system, where the increased surface area of the smaller feeder roots promotes the capillary action along the perimeter. However, when the lymph capillaries draw in fluid, they not only rely on molecule cohesiveness and molecule adherence to the vessel walls. Additionally, the restricted space inside of each capillary works to enhance the suction around the sodium proteins contained within and this rise in pressure, in turn collectively influences the direction of the interstitial fluid flow.

Once the fluid is drawn into the lymphatic capillaries, the ever-increasing vessel architecture creates additional space for the evenly distributed sodium protein concentrations to increase proportionately, within the expanding diameter of each vessel. Creating ionic pull through the system that draws the lymph fluid containing the contaminated sodium, in the correct direction. Permitting the lymphatic system to not rely solely on gravity, muscle contraction during movement, and breathing contractions in order to promote the lymphatic circulation pump.

The sodium protein transport system may be of particular use during periods of sleep, where the brain is washed clean of its associated waste by the cerebrospinal fluid. Since most of humanity sleeps in a still, prone position, according to present thinking, only the breathing mechanism is available to aid the lymphatics throughout the majority of the night.

This is not an efficient or safe strategy for the body to leave the bulk of the days brain waste suspended in the interstitial fluid, during the resting state. While removing this waste during the waking state, when most of the body systems are actively producing more waste. Logic dictates the body should remove this waste in a consistent manner, and ionic pull provides a mechanism for such around the clock removal.

Hypothetically, the sodium protein chains, residing in the interstitial fluid adjacent the lymphatic system, would effectively bridge the physical divide between the lymphatics and all other body systems. And then utilizing ionic pull, these protein chains could transport the fluid, along with any unwanted material to the lymph system and then through the vessels, in one continuous frequency.

Such a mechanism would permanently expand the physical boundaries of the lymph system to reach the brain where there are no lymph nodes, or temporarily expand these boundaries in the case of an unexpected toxic event, without the body having to devote any more space to the lymphatic structure.

Therefore, during an infection or toxic overload event, a linear organization of sodium protein chains, which have established increasing density within the interstitial fluid, may appear on imaging as telescopic channels. These channels would grow denser and/or wider in the direction of the flow, to remove this waste overload. If this is so, then early detection of any abnormalities in density or width of such sodium protein chain formations within the interstitial fluid, as compared to the baseline mapping of healthy individuals, could prove useful for diagnostic purposes.

Over the course of time, failed attempts by the lymph system to remove this excess waste from the body may cause a toxic buildup to occur within the corresponding lymph node region, which exceeds the ionic capacity of the node. In this case, one might predict, the sodium protein concentrations within the lymph node will elevate to levels higher than what would be previously determined by the baseline mapping of healthy individuals. These higher concentrations are necessary to maintain optimum ionic capacity under a heavy work load, and again might be an early marker for potential disease.

The buildup of sodium protein concentrations may first occur within the lymph node region and then incrementally on down the length of the lymph system, in the form of a linear channel, to encourage the trapped waste to continue to move in the right direction. To put it another way, once the sodium protein levels become elevated in a lymph node region, the corresponding channel downstream, must be proportionately increased as well, to maintain the ionic pull through the lymph system. If this is so, tracing the elevated protein concentrations within the lymph system, back to the original lymph node region displaying abnormally high concentrations, may be useful not only for revealing the location of the initial toxic buildup, but also for identifying any other sites effected by the pending disease.

When the sodium proteins within a lymph node are raised to their maximum level, and have reached full ionic capacity, the node itself will malfunction and no longer be able to properly neutralize toxins. A concern beyond the obvious would be the one-way valves, which encourage the lymph fluid to flow through the node in one direction, may begin to fail, due to the excessive trapped material. To prevent this leak from becoming life threatening, one might expect the lymphatic system to create a secondary, although less satisfactory outlet for the toxic lymph and/or interstitial fluid.

For this scenario, where the one-way valves and lymph node are underperforming, such diminished ionic pull through the node, would create an area of slow-flow or in a worst-case scenario no-flow just outside/upstream of the lymph node region. This backup of contaminated fluid occurring within the vessels, and/or outside of the vessel capillaries, would concentrate the localized contaminated sodium into ever higher levels.

In response, the theory predicts the possibility of the formation of sodium protein chains upstream from the compromised lymph node region, to clear away the surplus sodium. These sodium protein chains would develop ionic pull away from the damaged valves, in the opposite direction of the lymph system flow. Allowing the protein chains to siphon off and then deposit the stagnating fluid into the surrounding tissue, as a means to clear away the excessive waste material. In other words, these newly formed sodium protein chains will act as an emergency escape valve, temporarily bypassing the node and depositing all unfiltered fluid back into the local tissue.

Over time, if this condition persists and the surrounding tissue becomes saturated with the contamination, the low concentrations of sodium protein chains will most likely densify into higher concentrations, in an attempt to increase the pump action away from the lymph system. If this reaction does not clear the node, the sodium protein chains may then widen in diameter, in the direction of the siphon flow. And as a final stage, the chains could form many thick, dense branches, developing off of the original sodium protein chains, to broadly re-distribute the interstitial/lymph fluid. We should be able to utilize the images of these localized sodium protein chain signatures, to predict the duration of the node malfunction and the extent of the disease.

Such a deliberate bypass strategy, would temporarily reduce the incoming material from entering the lymph node, allowing the node and valve time to clear itself, recover and eventually be able to return to normal function. In the early stages, the advantage to this strategy is these sodium protein escape valves need only to consist of relatively low protein concentrations, to overcome the weaker capillary action of the low-flow lymph fluid. So, if the lymph node can clear itself and once again achieve its original ionic pull, the stagnant pool will be drawn back to the lymph system via normal uni-flow, effectively overriding the opposite pull of the sodium protein escape valves.

However, if the lymph node function cannot be restored and the lymph valve continues to leak, then the low-flow segment outside/upstream of the lymph node, will remain chronic. Forcing the sodium protein concentrations within the escape valve to increasingly higher levels, to pump further into the surrounding tissue, in order to distribute the chronic waste. This in turn creates stronger ionic pull away from the lymph system, to span the distance. Unfortunately, this augmented pull in the wrong direction, will also require proportionate increases in the sodium protein concentrations within the vessels themselves, if the corresponding channel is ever to re-establish the appropriate directional flow.

To circumvent such a life altering chain reaction, the body first chooses to dump the untreated or partially treated waste into the surrounding adipose fat. The reason being that fat is an ideal temporary storage tissue for disease and toxins, because when the toxic fat is burned for energy, it re-releases the stored toxin back into the bloodstream. This phenomenon temporarily relieves the overwhelmed node from the localized high concentrations of toxins, and hopefully re-distributes the burden of clean up elsewhere to less effected nodes, as the waste moves through the bloodstream and is filtered again. In this way, the lymphatic system can temporarily defer the cleanup and even change which nodes it will use for removal, while reducing the localized waste and the immediate need for excessive sodium protein buildup.

Since the endgame for the lymphatic system is to trap and neutralize excess toxins for the eventual removal by the kidney/spleen/liver, during a chronic situation one may expect to find excessive amounts of sodium in these organs, and in many cases this is true. The Lymphatic Frequency Theory predicts a gradual and incremental buildup of sodium proteins within these organs, too.

Such sodium protein buildup occurs when the ionically compromised sodium proteins in the lymph vessels, can no longer pump. As a result, they fill with fluid, which renders them unable to remain bonded within the chain and they slough off. Now designated as waste, they too are removed from the process, to make room for new sodium proteins to form. Regrettably, during chronic conditions, the buildup of spent sodium and sodium proteins, eventually compromises the waste organs ability to function, along with interrupting the lymphatic flow. This failure of the lymph system, where the two-beat flow of the lymph pump is compromised, may explain why fluid buildup in the lower extremities of patients with circulatory failure crescendos late in the day, because the majority of the day's waste is no longer processable in a timely manner and back flow occurs.

In essence, waste removal deferment is a deliberate and successful survival strategy, evolved over millenniums. Still, this mechanism has repercussions and may play a major role in the prevalence of breast cancer formation in females. Especially, when we consider most female breasts contain relatively large amounts of adipose fat, along with the fact that the brain, the nasal cavities, the ear canals and the oral cavity are all located upstream. So, when a node develops a leaky valve near the breast and the newly formed escape valve builds up excessive ionic pull, the theory suggests the upstream interstitial waste may be diverted away from the affected node and into the breast tissue.

Moreover, the stagnating lymph fluid located within a corresponding channel of the lymph system, directly downstream from the effected node, may indeed reverse flow as well. Allowing the lymph system to utilize the abundance of adipose fat tissue in the breast, as a chronic dumping ground, in an effort to clear the waste by diverting from above and/or backflowing from below the failing node. If a chronic situation develops, the breast tissue adjacent the escape valve will store ever-increasing amounts of waste.

In many instances, a female will burn the adipose tissue in the other more active areas of the body first, such as the trunk and legs. So conceivably, a situation might arise where an excessive amount of toxic lymph/interstitial fluid waste is deposited within the breast tissue, but not returned in a timely manner to the circulatory system via fat burning, eventually leading to breast cancer.

With this concept in mind, another important point to consider is that lymph nodes would not be optimal for filtering waste from the brain, because the brain is the fattiest organ in the body, which makes the brain a prime candidate for reverse flow waste dumping. Of equal concern is not only the dumping of the untreated waste, but also the partially treated lymph waste that is broken down into unfamiliar and as yet undetectable forms, through standard medical testing.

In theory, the back flowing of this partially broken-down undetected waste, may help shed light on neurological disorders like Multiple Sclerosis and Parkinson's disease, since the nervous system relies heavily on sodium proteins as a way to transport nerve impulses. If a compromised lymph node region adjacent the nervous system becomes chronic, the above-mentioned escape valve strategy may eventually extend itself back to the nervous system. This short-circuiting, would allow the lymphatic system to utilize the entire nervous system's sodium transport mechanism, for the purposes of redistributing the stagnating, contaminated fluid away from the compromised lymph node region, in an effort to clear the corresponding nodes.

In other words, the lymphatic system may reach a point where its mighty ionic capacity is diminished, so at least a channel within a lymph vessel will have less ionic pull then the nervous system, and reverse flow is established. The advantage here is the nervous system could transport the toxins to the surrounding lymph nodes still in working condition, to remove the waste overload.

Indeed, the aches and pain we all experience from an air-borne flu, may be a direct result of such a redistribution of infection. Although this transfer process is unpleasant to the patient, the symptoms are usually temporary, and the virus in most instances, is successfully removed without repercussion.

If, however the nodes surrounding the nervous system have become compromised, where the neighboring lymph vessels are no longer able to pull the toxins back out of the nervous system, than the sodium proteins within the nervous system itself, will reach ionic capacity. Thereafter, impairing the sodium proteins ability to transport the nerve impulses, resulting in the neurological symptoms displayed by such diseases.

Certainly, if sodium protein chains are utilized by the lymphatic system for transport, it should be possible to manipulate these sodium proteins locally and/or systemically, to exploit them for therapy. In addition, healthy adipose fat tissue could be harvested and reproduced in large numbers, via current stem cell technology. Subsequently, once we have identified the compromised lymph nodes, we could synthesize our own sodium protein escape valves and adipose fat collection sites, using marked material, which could be easily injectable and removable through manual lipo-siphoning practices, performed under MRI /ultrasound.

Such intervention will first and foremost, physically remove the localized toxin/infection, and prevent the contamination from further compromising the surrounding tissue. Removing the toxic overload, should break the cycle of sodium protein chain buildup and take the pressure off the lymphatic system and organs. Permitting the sodium protein concentrations to return to their established healthy baseline through attrition, and eventually normalizing the lymphatic function.

To reverse the process of later stage waste dumping, particularly in the female breast region, we may not only need to create synthetic escape valves and collection sites to remove the toxin. But in a worst-case scenario, would also need to reduce the concentrations of the existing sodium protein escape valves, in order to re-direct the localized flow of the syphoned lymph fluid.

This could be accomplished locally though injection, by saturating the existing proteins with an alkaline, ionic slurry such as baking soda.

The goal of injecting the existing escape valves, is to neutralize their ionic pull, by overwhelming the ionic capacity of the existing sodium protein chains, causing them to fill with fluid and then to slough off. Consequently, with the high concentration escape valves interrupted, the newly employed low concentration synthetic escape valves, will then have enough ionic pull to remove these individual spent proteins, along with the ability to siphon off the low-flow lymph fluid, without competition.

In the case of the nervous system, these same techniques would redirect the chronic re-circulating waste away from the nervous system and allow the nerves themselves time regenerate without toxic interference. Again, such intervention, should reduce the need for excessive ionic pull through the lymph system, and will eventually allow the protein concentrations within the lymph system itself, to normalize.

For lymph edema, where a corresponding lymph node region is either compromised, or is actually surgically removed, the flow is interrupted and a directional change occurs. We may be able to establish a new collection site adjacent the edema, to siphon off the low-flow lymph fluid. This will eventually lower the waste concentrations of the infected area, and allow the directional flow to normalize, or in a worst-case scenario where flow is surgically interrupted, the siphoning procedure could be continued indefinitely to minimize the edema.

Many dermatological disorders could be viewed in a similar fashion, where the lymphatic system is attempting to back flow through the pores of the skin, so than a chronic toxic exposure is irritating the surrounding skin tissue. These symptoms could be relieved in the same manner because the synthetic escape valves and collection sites are malleable and can be locally employed on an as needed basis.

Blood cancer diseases may be nothing more than a transfer problem from the circulatory system to the lymphatic system and might be circumvented by temporarily boosting the sodium, potassium and sodium protein levels in the interstitial fluid, to aid in waste transference.

Circulation issues could actually be a transfer problem as well, where the plaque is not properly removed from the blood stream, via the interstitial fluid. So eventually, late stage cardiac disease patients will present with compromised waste organs that are overwhelmed with spent sodium and sodium proteins, having lost their ionic capacity, while attempting to clear the blood vessels and the heart of plaque buildup. Sodium protein manipulation within the interstitial fluid, along with synthetic escape valves should work here, as well.

Rheumatoid arthritis may be the result of the lymph system dumping chronic waste into the soft tissue of the joints, rather than into the fat, via the nervous system. Additionally, most autoimmune diseases might be the body attempting to cordon off the end result of this waste dumping strategy. In actuality, the immune system is attacking the deposited toxins at a microscopic level, and must compromise the intermediate and healthy tissue, directly adjacent the infected tissue. Scleroderma, may merely be a tissue repair overreaction, after this process initiates and becomes chronic.

If so, then suppressing the immune system as a treatment for autoimmune disease is merely reducing the symptoms, which exist as collateral damage, but does not address the underlying disease. The ironic twist to autoimmune disease is quite possibly, the only way to fully recover from this scenario, may be to facilitate the immune system's effort to clear the waste, and not to suppress it.

Secondary infections such as bacterial and fungal infections, which are commonly found in association with chronic disease, may proliferate simply because there is not enough ionic capacity within the system, to pull the unwanted organisms through the neutralization process in a timely manner. Therefore, the partially treated infection is able to overpopulate and then is dumped unchecked into a specific site. By restoring the ionic capacity of the lymph system through therapy, we should be able to minimize these secondary infections, along with the chronic disease itself, and we may find the two coexisting disorders inextricably linked.

Certain forms of the Herpes virus, which resides in the brain, take advantage of the lack of lymph node protection for the brain, and utilize the sodium proteins of the nervous system when active. Herpes could be particularly vulnerable to synthetic escape valve collection sites, strategically placed to divert the Herpes away from the nervous system. These collection sites could be put in place ahead of an outbreak, and then maintained indefinitely, until the virus is fully captured. Or at least until the virus is minimized to a point where the immune system can eradicate the issue on its own.

A multitude of brain disorders, including some forms of childhood autism, may be explained by the Lymphatic Frequency Theory. Where the aluminum, antibiotics, formaldehyde, human aborted fetal apparatus, MSG and mercury from numerous vaccines administered simultaneously, overwhelms the developing lymph system, causing toxic dumping to occur in the brain. We could strategically raise the interstitial sodium protein levels in high risk cases, prior to administering the vaccine combination, as a preemptive measure or even after the vaccine if the need arises. Genetic disorders such as Cystic Fibrosis, may be aided at least symptomatically, by increasing the sodium protein levels in the interstitial tissue adjacent the lungs, to help remove the excess sticky cells, which clog the patient's airways. Permitting the patient to better co-exist with the disease and extending their life span.

Interestingly enough, under this theory genetic markers that predict diseases would garner a different meaning, for successful genetics, evolved to promote survival strategies over millenniums, are now inadequate. Not due to faulty genes, but to a changing environment, where some of today's toxins and pathogens, can no longer be successfully cleared from the body. Yet, deferment of waste removal by the lymphatic system is a critical component to our survival and is not something we can avoid by changing our genes, for toxic overload affects us all, regardless of our genetic makeup.

In essence, gene manipulation to avoid a perceived body deficiency, could merely be trading in one chronic disease for another, because toxic overload cleanup through deferment occurs in every body type. Therefore, the main issue we face, is understanding where the genes are telling the body to dump the chronic waste. In this way, genetic markers may help us to predict which chronic waste deferment strategy, is prevalent within each body type. Such knowledge would allow us to standardize the wellness monitoring of the sodium proteins and could be customized for each individual.

In addition, we could analyze the various toxins and pathogens, to further correlate each body type reaction, to each specific contamination. With an understanding that the body may have more than one way to defer the cleanup of a toxic overload, depending on the chemical makeup of the actual waste or infection. Once we fully comprehend these dynamics, we can develop specific sodium protein protocols, in order to front run the most typical diseases in each genetic category. Rather than waiting until a disease is full blown, and then trying to react to the fallout.

Unfortunately, modern medicine attempts to combat disease in the exact opposite manner. Where drugs are commonly symptom based, locking a patient into chronic dependency, and further overloading the already malfunctioning sodium protein transport system. Consequently, such medical practices are promoting the disease and generating additional symptoms, requiring more drugs that continue the patient's downward spiral.

We can and should do better for ourselves. Instead of treating a disease in retrospect, sodium protein monitoring will place intervention ahead of the disease. Accordingly, the next generation of drugs will be designed to enhance the deferment strategy of the lymphatic system, not defeat it. To have success, we must learn to target medical intervention towards manipulating the sodium protein action within the interstitial transition zone, the capillary entry points of the lymphatic system, the nodes, vessels and the exit points or the waste organs, to aid the natural clearing process.

The pharmaceutical industry should be developing drugs to temporarily elevate and/or reduce the lymphatic sodium proteins, at each specific point within this process. We should also be able to manipulate the enzymes controlling the sodium and potassium production levels within this system, so the two remain proportionate. Thus, promoting the optimum level of ion exchange, to maximize the sodium protein pumping action.

Furthermore, we could expedite waste elimination by synthetically encouraging the early removal of any partially spent sodium proteins, within or around the lymph system, to make room for new proteins. Provided, we can help the organs to clear such an accelerated waste neutralization process. To successfully enhance this accelerated clearing process, we should be able to create synthetic escape valves and collection sites upstream from the waste clearing organs, which will temporarily relieve the overwhelmed organs from the onslaught of spent sodium and sodium proteins.

Again, these escape valves would consist of low concentration sodium protein chains, which draw from the slow-flow lymph fluid outside of the corresponding node region. Once the problem is cleared enough, the lymphatic uni-flow will re-establish, and naturally bypass the synthetic escape valves. Leaving the escape valves and collection sites intact for future monitoring and use.

Alternative medicine can also play a role in aiding the lymph system with waste removal. Charcoal and sphagnum peat moss can both be consumed orally in pill form, because they retain much of their ionic capabilities beyond the stomach. Such binding properties help to clear the intestine and liver of chronically recycled waste and should be considered as a first line defense tool to complement sodium protein therapy. Orally consumed milk thistle, and many other plants can aid the function of an overwhelmed liver. A charcoal/ clay mask applied to the skin, may help to draw out toxins trapped in the skin. Epsom salt baths can ionically remove toxins through the pores and the bonding action of salt water mist can help the lungs clear particulate matter.

Henceforth, a paradigm shift needs to occur in modern medicine, where the focus is placed on the transportation of waste through the various stages found within a malfunctioning lymphatic system, including the transference of waste through the interstitial region. With the understanding that such waste transport break downs, are absolute precursors to predicting the failure of the other body systems.

Keeping this in mind, the protocol for sodium protein therapy would include, but not be limited to; sodium protein wellness monitoring and diagnostics for each specific body type, physical removal of toxins through constructing localized synthetic escape valves and collection sites, physical interruption and removal of the existing escape valves using an alkaline slurry injected into the local site, synthetic sodium protein and enzyme manipulation within and adjacent the lymph system and other body systems. And finally, toxic waste removal using alternative supplements.

In this way, we may be able to address a multitude of body malfunctions, by focusing on the commonality of their underlying cause, which is a toxic waste overload within or adjacent the lymphatic system. Allowing us to intervene long before the chain-reaction breakdown of the lymphatic system occurs, where the end result of such a systematic failure will manifest elsewhere, as chronic disease.

Modern medicine has largely ignored the fundamentals of toxic waste removal management. Creating an environment where the patient is subjected to a symptom based, helter-skelter approach that often times places zero consideration on lymph system failure. As a result, many of today's medical treatments are masking symptoms, while actually furthering the progression of the disease. In this century, if we so chose to monitor and combat disease at the lymphatic transport level, we stand a good chance of mitigating or eliminating the vast majority of illnesses that currently plague humanity.

About the Author

Ken Liskey attended the University of Delaware where he received a B.S. degree in Plant Science, with a minor in Landscape Design. He was partners in a Landscape Design/Build firm for 30 years, where the firm won numerous awards exhibiting at the Philadelphia Flower Show, including Best in Show.

Ken holds a patent with his father for inventing an apparatus, which recycles the air from the surface of a moving vehicle into usable electricity, while lowering the rear end drag as an interactive vortex generator.

Ken's diverse background of formal training, lifelong connection to the earth and invention experimentation, provoked him to write a soon to be released novel, portraying universal energy as a continuous frequency. His upcoming book, the **Shaman of Dark Matters**, describes energy in a constant state of transition between the physical realm and the spiritual realm.

This notion spawned the idea that the transference of waste from the various body systems to the lymph system, could be viewed as a continuous frequency, as well. Prompting Ken to model his Lymphatic Frequency Theory after his Universal Frequency Theory. Where a fundamental underlying current must

exist, not only when transferring energy from one realm to the other within our universe, but also when transporting waste from one system to another within the body.

Made in United States
Troutdale, OR
11/03/2026

56844421R00022